HELLO, BODY!

MUSCLES

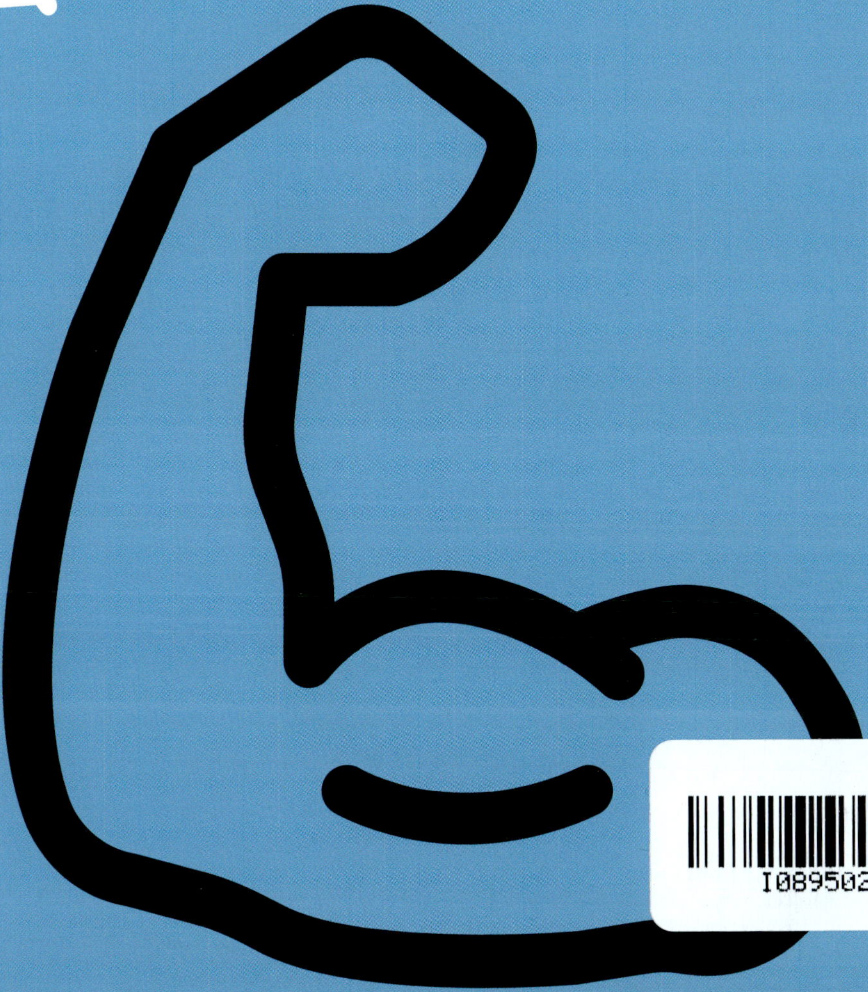

by Joyce Markovics

Cherry Lake Press
Ann Arbor, Michigan

Cherry Lake Press

Published in the United States of America by Cherry Lake Publishing Group
Ann Arbor, Michigan
www.cherrylakepublishing.com

Reading Adviser: Beth Walker Gambro, MS Ed., Reading Consultant, Yorkville, IL
Content Advisers: Sharon Markovics, MD, and Peter Markovics, MD
Book Designer: Ed Morgan

Photo Credits: freepik.com, 4; Artur Andrzej/Wikimedia Commons, 4 bottom; Artur Andrzej/Wikimedia Commons, 5; Wikimedia Commons, 6; Wikimedia Commons, 7; freepik.com, 8; © Jolygon/Shutterstock, 9; freepik.com, 10; freepik.com, 11; freepik.com, 12; freepik.com, 13; freepik.com, 14; freepik.com, 15; freepik.com, 16; © madespix/Shutterstock, 17; freepik.com, 18; freepik.com, 19; freepik.com, 20–21.

Cherry Lake Press is an imprint of Cherry Lake Publishing Group.

Library of Congress Cataloging-in-Publication Data

Names: Markovics, Joyce L., author.
Title: Muscles / by Joyce Markovics.
Description: Ann Arbor, Michigan : Cherry Lake Publishing, [2023] | Series: Hello, body! | Includes bibliographical references and index. | Audience: Grades 4-6
Identifiers: LCCN 2022003843 (print) | LCCN 2022003844 (ebook) | ISBN 9781668909638 (hardcover) | ISBN 9781668911235 (paperback) | ISBN 9781668914410 (pdf) | ISBN 9781668912829 (ebook)
Subjects: LCSH: Muscles–Juvenile literature.
Classification: LCC QP321 .M3325 2023 (print) | LCC QP321 (ebook) | DDC 612.7/4–dc23/eng/20220330
LC record available at https://lccn.loc.gov/2022003843
LC ebook record available at https://lccn.loc.gov/2022003844

Printed in the United States of America by
Corporate Graphics

CONTENTS

HELLO, BODY!

STRONG WOMEN

Can you imagine crushing a frying pan with your bare hands? How about lifting an adult over your head? Aneta Florczyk can. The Polish athlete might not look like one of the strongest people in the world, but she is.

Aneta Florczyk

At the age of 16, Aneta starting lifting weights for fun. And she never stopped. As an adult, she competed against other weightlifters. For several years, Aneta won the title of world's strongest woman!

During a competition, Aneta lifted 12 adult men, one at a time, over her head!

Incredibly, many other strong women came before Aneta. Starting in the late 1800s, women performed **feats** of strength in shows and circuses. In 1911, a reporter saw the great Sandwina use her husband as a human **barbell**. Then Sandwina "twirled her husband about in dizzy circles above her head," wrote the reporter. Finally, Sandwina tossed him into the air!

Strong women like Sandwina juggled adult humans, lifted hundreds of pounds, and bent steel bars. They carried out all of these amazing feats with their muscles!

WHAT ARE MUSCLES?

It's not just weightlifting that requires muscles. Touch your nose. Point your toes. Swallow and then laugh. All of these movements require your remarkable muscles. Even when you're sitting perfectly still, your muscles are hard at work.

There are two main categories of muscle: **voluntary** and **involuntary**. You can control voluntary muscles. However, involuntary muscles are controlled by your brain. Without thinking about them, muscles make your heart beat. And they help your chest rise and fall as you breathe. Muscles also help push blood through your blood vessels.

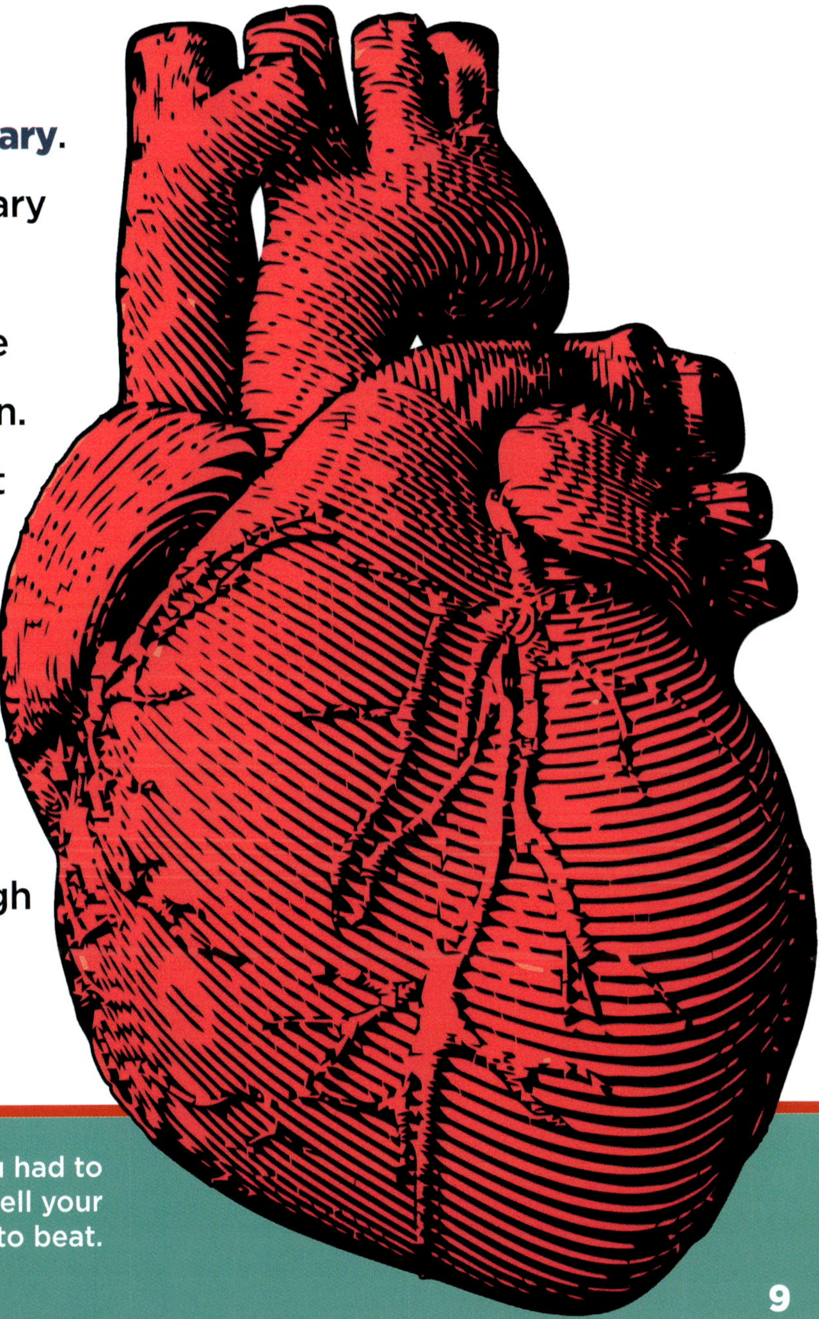

Imagine if you had to remember to tell your heart to beat.

There are more than 600 muscles in your body! They're everywhere—from your fingers and scalp to inside your body. Yet all muscles are alike in that they contract, or shorten. When a muscle contracts, it moves the body part to which its attached.

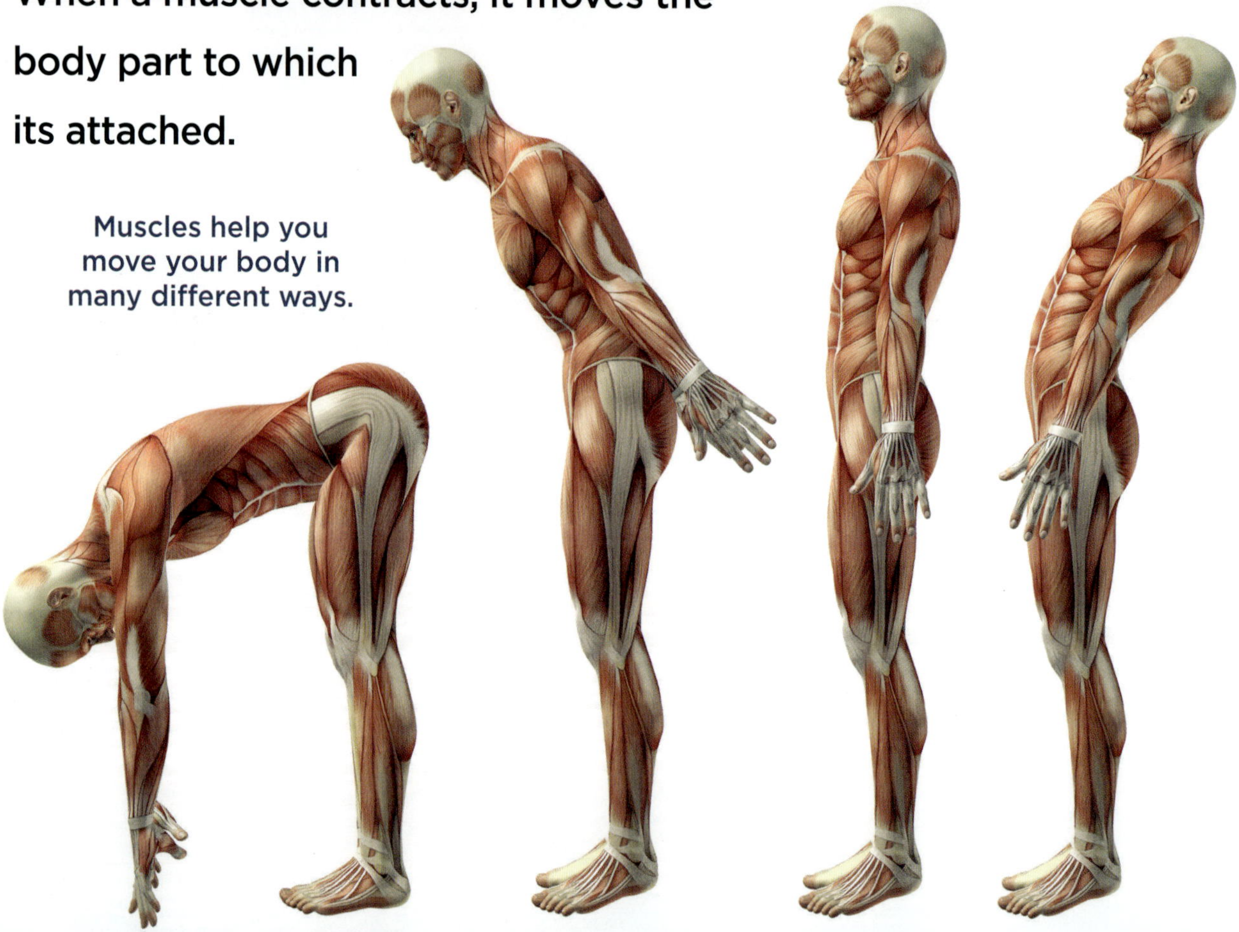

Muscles help you move your body in many different ways.

All muscles are made from the same type of material—a bundle of **fibrous**, stretchy **tissue**. Thousands of super-tiny fibers make up each muscle. Think of your muscles as giant rubber bands made from lots of smaller rubber bands.

The largest muscle in the human body is the gluteus maximus (GLOO-tee-uhs MAK-suh-muhs). It's in your backside! This muscle is large and powerful to help you stand upright and climb stairs, for example.

MUSCLE TYPES

You have three different types of muscles in your body. They are cardiac (KAR-dee-ak), smooth, and skeletal (SKEH-luh-tuhl) muscles. Together, these muscles make up your muscular system. The walls of the heart are made from cardiac muscle. This muscle is responsible for your heart pumping blood to your body. It's an involuntary muscle.

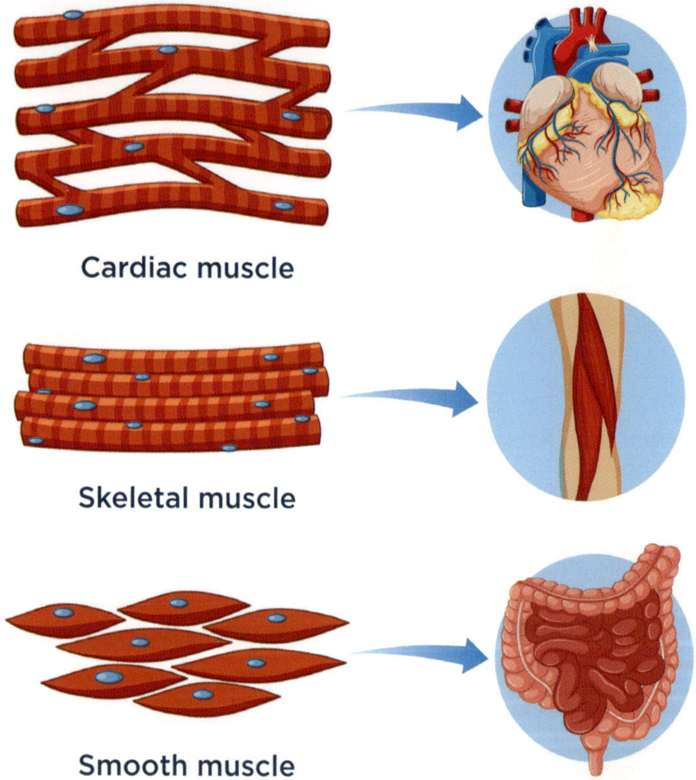

Smooth muscle is another muscle type we can't control. Its fibers look smooth under a **microscope**. Smooth muscle is found almost everywhere in your body. For example, it lines the stomach and intestines.

Smooth muscle helps you vomit when you're sick. With great force, it pushes food from the stomach out of the mouth. *Yuck!*

Smooth muscle is also found in your blood vessels. The muscle directs the movement of blood by contracting and relaxing. This helps to control **blood pressure**. You can also find smooth muscles in the bladder and uterus. In the bladder, as the muscle relaxes, it helps you hold in **urine**. Then it contracts, so you can push out urine when you pee!

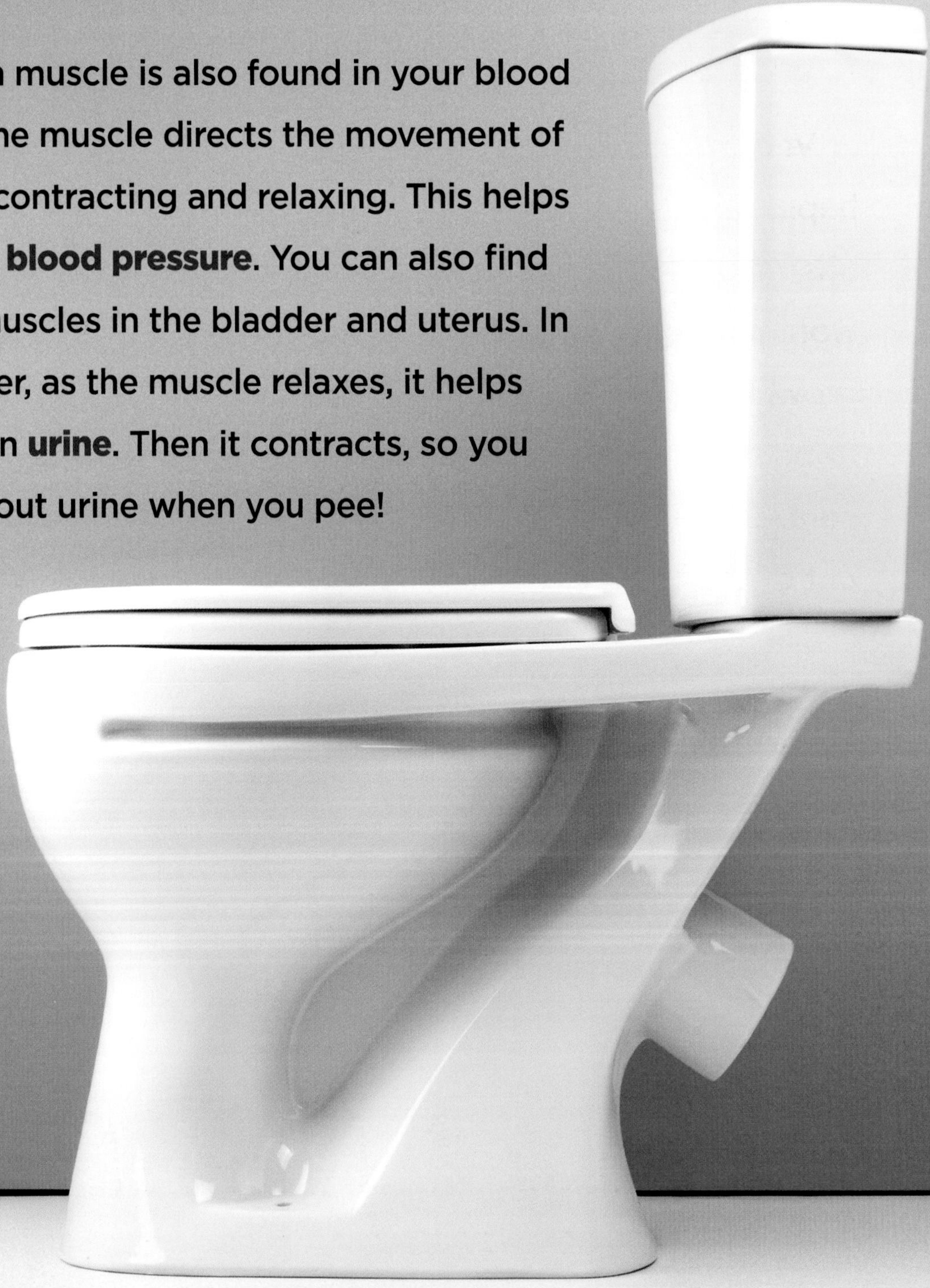

Very strong muscle **contractions** happen in the uterus (YOO-ter-uhs). This is the organ in a woman's body where babies grow. When a baby is ready to be born, the muscle contractions push the baby out of the mother. Go, Mom!

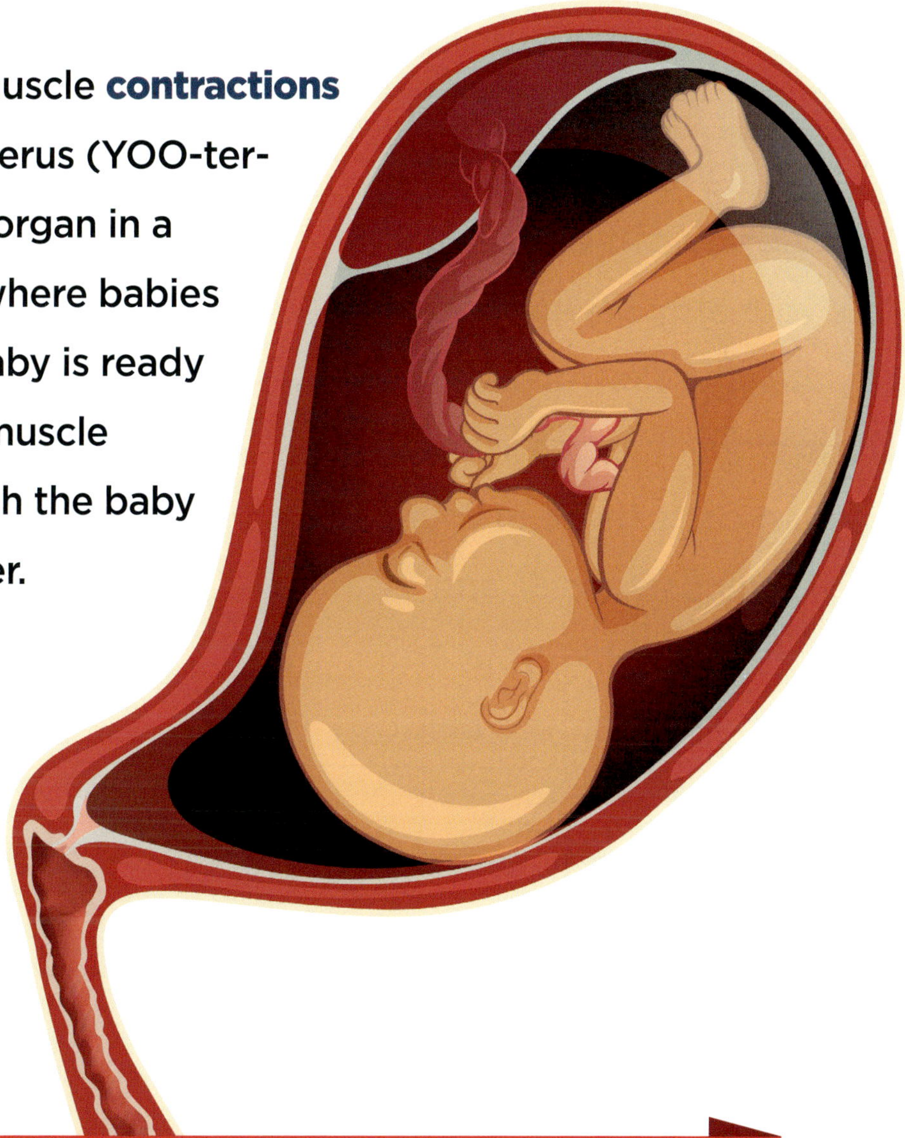

There are smooth muscles in inside your eyes too. They help you focus on whatever you're looking at.

The final type of muscle is the one you can see when you flex your arm. It's known as skeletal muscle. You control all the skeletal muscles in your body, such as those in your legs, shoulders, neck, and arms. For example, your arm won't bend to throw a ball unless you want it to. Skeletal muscles contract quickly and are some of the strongest in the body.

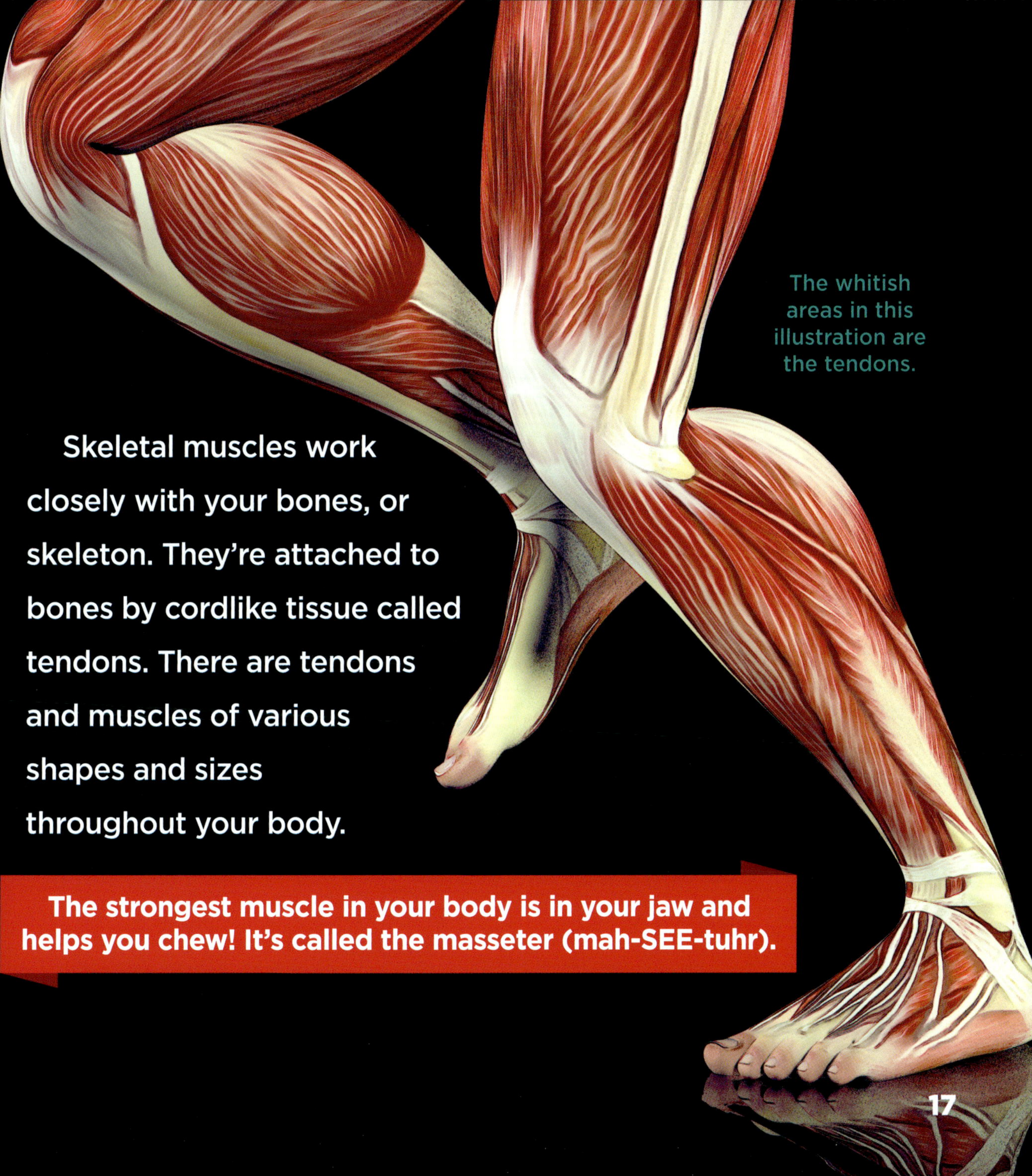

The whitish areas in this illustration are the tendons.

Skeletal muscles work closely with your bones, or skeleton. They're attached to bones by cordlike tissue called tendons. There are tendons and muscles of various shapes and sizes throughout your body.

The strongest muscle in your body is in your jaw and helps you chew! It's called the masseter (mah-SEE-tuhr).

FACE AND TONGUE

Skeletal muscles attach to the bones and skin in your face. Your face alone has more than 40 muscles. Take a look in the mirror and make a bunch of funny faces. You can contract your facial muscles a lot or a tiny bit. It's only a slight movement that can turn a frown into a smile!

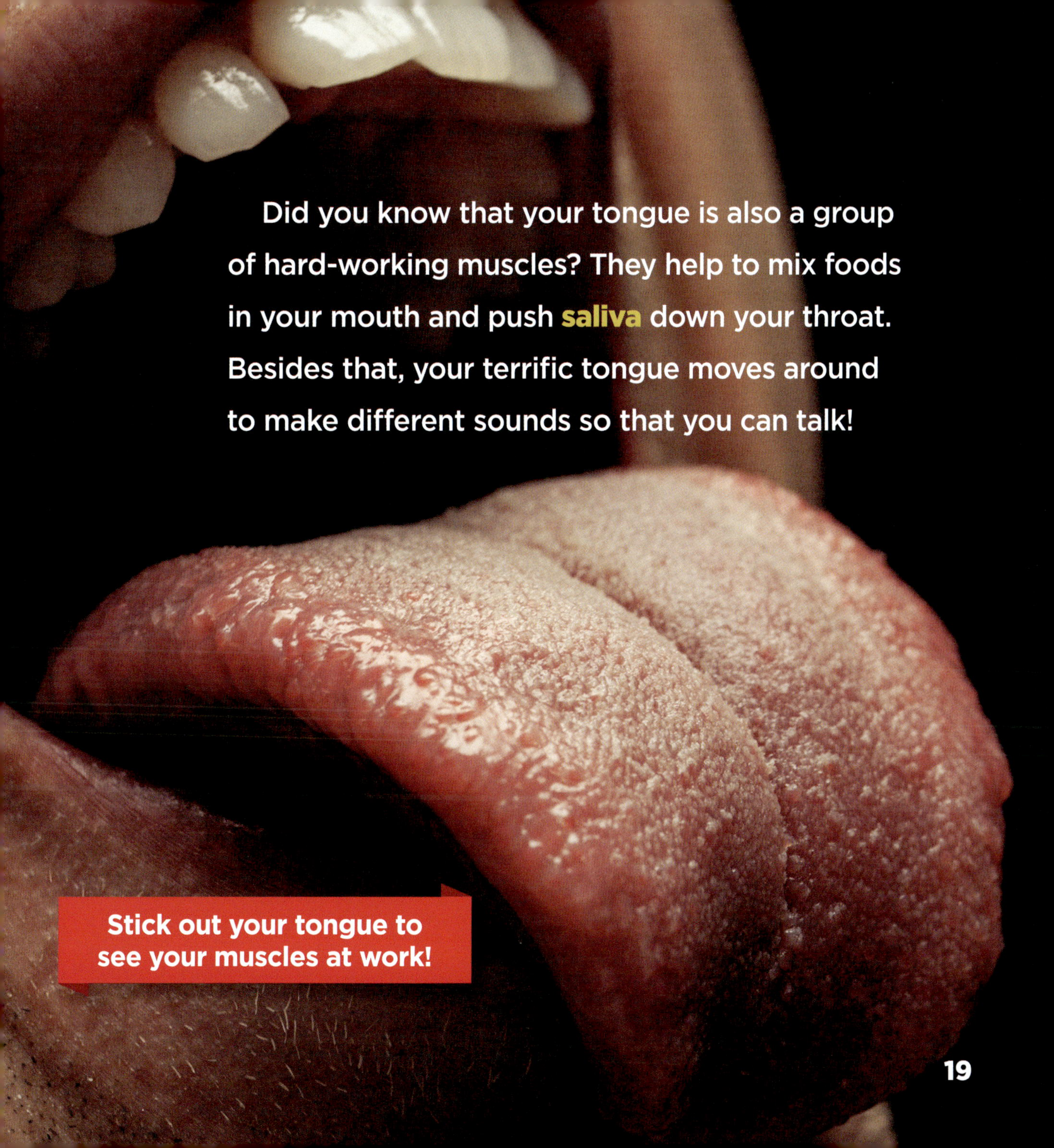

Did you know that your tongue is also a group of hard-working muscles? They help to mix foods in your mouth and push **saliva** down your throat. Besides that, your terrific tongue moves around to make different sounds so that you can talk!

Stick out your tongue to see your muscles at work!

MUSCLE MAYHEM

There are many diseases and injuries that can affect a person's muscles. Muscular dystrophy weakens the muscles over time. Because this disease affects all the muscles in the body, some people have heart, walking, and breathing problems.

Much more common are muscle injures. They include bruises, strains, and **sprains**. Have you ever strained or pulled a muscle? That happens when you overstretch a muscle or tendon. Thanks to your amazing body, minor injuries often heal on their own!

HEALTH TIPS

Here are some ways to keep your muscles healthy:

- Be active! Exercise at least three times a week. Walk, play, bike, jump, or dance.
- Maintain a healthy weight. Eat lots of healthy foods, such as vegetables, fruits, and whole grains. And avoid sugary drinks such as soda and fruit juice.
- Drink plenty of water!

GLOSSARY

barbell (BAHR-bell) a long metal bar with weights attached at each end

blood pressure (BLUHD PREH-shuhr) a measurement of the health of the heart and circulatory system

contractions (kuhn-TRAK-shuhns) the shortening of muscles in the uterus before and during childbirth

feats (FEETS) achievements that require skill or strength

fibrous (FYE-bruhs) consisting of fibers

involuntary (in-VAH-luhn-ter-ee) not by one's own choice

microscope (MYE-kruh-skohp) a tool used for viewing very small things

saliva (suh-LYE-vuh) a watery liquid in the mouth that helps with chewing and swallowing

sprains (SPRAYNZ) twists or stretches of the joints that cause pain and swelling

tissue (TISH-oo) a group of similar cells that forms a part of or an organ in the body

urine (YUHR-uhn) liquid waste matter made by the body

voluntary (VAH-luhn-ter-ee) by one's own choice

FIND OUT MORE

BOOKS

Johnson, Rebecca L. *Your Muscular System*. Minneapolis, MN: Lerner Publications, 2013.

Simon, Seymour. *The Human Body*. New York, NY: HarperCollins, 2008.

Simon, Seymour. *Muscles: Our Muscular System*. New York, NY: Morrow Junior Books, 1998.

WEBSITES

Britannica: Human Muscle System
https://www.britannica.com/science/human-muscle-system

Library of Congress: Muscles
https://www.loc.gov/everyday-mysteries/biology-and-human-anatomy/item/what-is-the-strongest-muscle-in-the-human-body/

Scholastic Study Jams! Muscular System
https://studyjams.scholastic.com/studyjams/jams/science/human-body/muscular-system.htm

INDEX

ABOUT THE AUTHOR

Joyce Markovics has written hundreds of books for kids. She marvels at the human body and how it can repair itself. She dedicates this book to her fellow Animal Nation volunteers—Priscilla, Summer, and Trish—three strong ladies with big hearts.